Tucholsky Wagner Zola Scott Sydow Freud Schlegel
Turgenev Wallace Fonatne
Twain Walther von der Vogelweide Fouqué Friedrich II. von Preußen
Weber Freiligrath Frey
Ernst
Fechner Fichte Weiße Rose von Fallersleben Kant Richthofen Frommel
Hölderlin
Engels Fielding Eichendorff Tacitus Dumas
Fehrs Faber Flaubert Eliasberg Ebner Eschenbach
Feuerbach Maximilian I. von Habsburg Fock Eliot Zweig Vergil
Ewald
Goethe Elisabeth von Österreich London
Mendelssohn Balzac Shakespeare Dostojewski Ganghofer
Lichtenberg Rathenau Doyle Gjellerup
Trackl Stevenson Hambruch
Mommsen Tolstoi Lenz Hanrieder Droste-Hülshoff
Thoma
Dach Verne von Arnim Hägele Hauff Humboldt
Reuter Rousseau Hagen Hauptmann Gautier
Karrillon Garschin
Defoe Hebbel Baudelaire
Damaschke Descartes Hegel Kussmaul Herder
Wolfram von Eschenbach Dickens Schopenhauer Rilke George
Bronner Darwin Melville Grimm Jerome Bebel Proust
Campe Horváth Aristoteles Federer
Bismarck Vigny Barlach Voltaire Herodot
Gengenbach Heine
Storm Casanova Tersteegen Grillparzer Georgy
Chamberlain Lessing Gilm
Brentano Langbein Gryphius
Claudius Schiller Lafontaine
Strachwitz Schilling Kralik Iffland Sokrates
Katharina II. von Rußland Bellamy
Gerstäcker Raabe Gibbon Tschechow
Löns Hesse Hoffmann Gogol Wilde Gleim Vulpius
Luther Heym Hofmannsthal Klee Hölty Morgenstern Goedicke
Roth Heyse Klopstock Kleist
Luxemburg Puschkin Homer Mörike Musil
La Roche Horaz
Machiavelli Kierkegaard Kraft Kraus
Navarra Aurel Musset Kind Moltke
Nestroy Marie de France Lamprecht Kirchhoff Hugo
Laotse Ipsen Liebknecht
Nietzsche Nansen
Marx Lassalle Gorki Klett Leibniz Ringelnatz
von Ossietzky
May vom Stein Lawrence Irving
Petalozzi Knigge
Platon Pückler Michelangelo Kafka
Sachs Poe Liebermann Kock
Korolenko
de Sade Praetorius Mistral Zetkin

The publishing house tradition has created the series **TRADITION CLASSICS**. It contains classical literature works from over two thousand years. Most of these titles have been out of print and off the bookstore shelves for decades.

The book series is intended to preserve the cultural legacy and to promote the timeless works of classical literature. As a reader of a **TRADITION CLASSICS** book, the reader supports the mission to save many of the amazing works of world literature from oblivion.

The symbol of **TRADITION CLASSICS** is Johannes Gutenberg (1400 – 1468), the inventor of movable type printing.

With the series, tradition intends to make thousands of international literature classics available in printed format again – worldwide.

All books are available at book retailers worldwide in paperback and in hardcover. For more information please visit: www.tredition.com

tredition was established in 2006 by Sandra Latusseck and Soenke Schulz. Based in Hamburg, Germany, tredition offers publishing solutions to authors and publishing houses, combined with worldwide distribution of printed and digital book content. tredition is uniquely positioned to enable authors and publishing houses to create books on their own terms and without conventional manufacturing risks.

For more information please visit: www.tredition.com

Drug Supplies in the American Revolution

George B. Griffenhagen

Imprint

This book is part of the TREDITION CLASSICS series.

Author: George B. Griffenhagen
Cover design: toepferschumann, Berlin (Germany)

Publisher: tredition GmbH, Hamburg (Germany)
ISBN: 978-3-8495-0456-4

www.tredition.com
www.tredition.de

Copyright:
The content of this book is sourced from the public domain.

The intention of the TREDITION CLASSICS series is to make world literature in the public domain available in printed format. Literary enthusiasts and organizations worldwide have scanned and digitally edited the original texts. tredition has subsequently formatted and redesigned the content into a modern reading layout. Therefore, we cannot guarantee the exact reproduction of the original format of a particular historic edition. Please also note that no modifications have been made to the spelling, therefore it may differ from the orthography used today.

DRUG SUPPLIES IN THE AMERICAN REVOLUTION

by George B. Griffenhagen

At the start of the Revolution, the Colonies were cut off from the source of their usual drug supply, England. A few drugs trickled through from the West Indies, but by 1776 there was an acute shortage.

Lack of coordination and transportation resulted in a scarcity of drugs for the army hospitals even while druggists in other areas resorted to advertising in order to sell their stocks. Some relief came from British prize ships captured by the American navy and privateers, but the chaotic condition of drug supply was not eased until the alliance with France early in 1778.

The Author: *George Griffenhagen — formerly curator of medical sciences, United States National Museum, Smithsonian Institution — is director of communications, American Pharmaceutical Association, and managing editor, Journal of the American Pharmaceutical Association.*

As one historian has reminded us, "few fields of history have been more intensively cultivated by successive generations of historians; few offer less reward in the shape of fresh facts or theories" than does the American Revolutionary War. [1] This is true to some extent even in the medical history of the Revolution. The details of the feud within the medical department of the army have been told and retold. [2] Even accounts of the drugs employed and pharmaceutical services have been presented, primarily in the form of biographies and as reviews of the *Lititz Pharmacopoeia* of 1778. [3] However, practically nothing has been published on the actual availability of medical supplies. Furthermore, the discovery of several significant but unrecorded account books of private druggists who furnished [Pg 111] sizable quantities of drugs to the Continental Army and a careful re-evaluation of the unusually significant papers [4] of Dr. Jonathan Potts, Revolutionary War surgeon, justify a review of the drug supplies during the early years of the war.

Continental Medicine Chests

As early as February 21, 1775, the Provincial Congress of Massachusetts appointed a committee to determine what medical supplies would be necessary should colonial troops be required to take the field. Three days later the Congress voted to "make an inquiry where fifteen doctor's chests can be got, and on what terms"; and on March 7 it directed the committee of supplies "to make a draft in favor of Doct. Joseph Warren and Doct. Benjamin Church, for five hundred pounds, lawful money, to enable them to purchase such articles for the provincial chests of medicine as cannot be got on credit." [5]

A unique ledger of the Greenleaf apothecary shop of Boston [6] reveals that this pharmacy on April 4, 1775, supplied at least 5 of the 15 chests of medicines. The account, in the amount of just over £247, is listed in the name of the Province of the Massachusetts Bay, and shows that £51 was paid in cash by Dr. Joseph Warren. The remaining £196 was not paid until August 10, after Warren had been killed in the Battle of Bunker Hill.

The 15 medicine chests, including presumably the five supplied by Greenleaf, were distributed on April 18 — three at Sudbury and two each at Concord, Groton, Mendon, Stow, Worcester, and Lancaster. [7] No record has been found to indicate whether or not the British discovered the medical chests at Concord, but, inasmuch as the patriots were warned of the British movement, it is very likely that the chests were among the supplies that were carried off and hidden. The British destroyed as much of the remainder as they could locate. [8]

Figure 1.—Medicine scales and oval box of medicinal herbs used by Dr. Solomon Drowne during the Revolution. Preserved at Fort Ticonderoga Museum, New York.

Two days after the battles at Lexington and Concord, the Provincial Congress ordered that a man and horse be made available to transport medicines. On April 30, Andrew Craigie was appointed to take care of these medical stores and deliver them as ordered.

Medical supplies were an early source of anxiety to the Provincial Congress of Massachusetts. The supply of drugs in Boston must have been largely controlled by the British after Lexington-Concord, and the limited supply in the neighboring smaller towns was soon exhausted. Four days before the Battle of Bunker Hill the Congress "Ordered that Doct. Whiting, Doct. Taylor and Mr. Parks, be a

7

committee to consider some method of supplying the several surgeons of the army with medicines," and further "Ordered that the same committee bring in a list of what medicines are in the medical store." [9]

On June 10 the responsibility of furnishing medical supplies to the army at Cambridge shifted to Philadelphia when the Continental Congress accepted the [Pg 112] request of the Massachusetts Provincial Congress to assume control and direction of the forces assembled around Boston. The Continental Congress established a Continental Hospital Plan on July 27, but it was not until September 14 that the Congress appointed a "committee to devise ways and means for supplying the Continental Army with medicines." On this same day, the deputy commissary general was directed to pay Dr. Samuel Stringer for the medicines he purchased, [10] which, as we learn later, were the initial supply for the Canadian campaign.

The first recorded purchase of drugs made directly by Congress, on September 23, was "a parcel of Drugs in the hands of Mr. Rapalje, which he offers at the prime cost." [11] Then, on November 10, Congress ordered that the medicine purchased in Philadelphia for the army at Cambridge be sent there by land. [12] But difficulties of supply commenced early. On January 1, 1776, Eliphalet Dyer wrote Joseph Trumbull asking "how could the cask of Rhubarb which was sent by order of Congress and was extremely wanted in the Hospital lye by to this time. After you came way I wrote to Daniel Brown to see it delivered." [13]

In the fall of 1775 there must have been a reasonably good stock of drugs in the hands of private Philadelphia druggists, and until the end of summer there were still a number of ships from Jamaica, Bermuda, Antigua, and Barbados putting in at Philadelphia with supplies, much of which originally came from England. Philadelphia druggists included William Drewet Smith, "Chemist and Druggist at Hippocrates's Head in Second Street"; [14] Dr. George Weed in Front Street; [15] Robert Bass, "Apothecary in Market-Street"; Dr. Anthony Yeldall "at his Medicinal Ware-House in Front-Street"; [16] and the firm of Sharp Delaney and William Smith. [17] The largest pharmacy in Philadelphia was operated by the Marshall brothers—Christopher Jr. and Charles. This pharmacy had been established in

1729 at Front and Chestnut Streets by Christopher Marshall, Sr., a patriot who took an active part in the care of the sick and wounded in Philadelphia hospitals during the Revolution. [18]

As the plans progressed for raising troops from New Jersey, Maryland, Delaware, Pennsylvania, Virginia, North Carolina, and South Carolina, Congress called on the committee on medicines "to procure proper medicine chests for the battalions...." [19] The journal of the Continental Congress fails to indicate the source of these medicine chests, but the Marshall brothers' manuscript "waste book" (daily record) for the period February 21 to July 6, 1776, [20] indicates that the Marshall apothecary shop was the primary supplier. The records show that the Marshalls furnished 20 medicine chests to the following battalions from February to June: [21]

February 1776:	Pennsylvania 1st Battalion
March 1776:	Jersey 3d Battalion
April 1776:	Pennsylvania 2d, 3d, and 6th Battalions
May 1776:	Six Virginia battalions Jersey 1st Battalion Pennsylvania 4th Battalion
June 1776:	Six North Carolina battalions Virginia 9th Battalion

The exact contents of each chest are indicated in the Marshalls' waste book. The chest furnished to the Pennsylvania 4th Battalion is an example of the ones supplied by Congress in the spring of 1776; its contents are listed on page 130.

Congress intended that all chests be substantially the same, but the amount of medicines demanded exceeded the stock of even the largest druggists. The first several chests were complete as ordered, but as early as April the Marshalls were running out of [Pg 113] certain drugs. Gum opium and nitre "found by Congress" was included in the chest for the Pennsylvania 4th Battalion, and by May 11 the Marshalls were out of Peruvian bark, ipecac, cream of tartar, gum camphor, and red precipitate of mercury. The chests outfitted after June 1 also failed to include Epsom salts, and the last chest lacked jalap as well. Thus the majority of the battalions traveling north were already without some of the most necessary drugs in their chests. Blithely their medical officers thought they could obtain the missing drugs when they arrived at the general hospital.

Treason, Poison, and Siege

After the Battle of Bunker Hill, the forces around Boston settled down for a 9-month siege. Two days after General Washington arrived in Cambridge on July 2, 1775, to take command of the army, the Provincial Congress of Massachusetts ordered a committee to prepare a letter informing him of the provisions that had been made for the sick and wounded of the army. On the very same day, July 4, the Provincial Congress appointed Andrew Craigie medical commissary and apothecary for the Massachusetts army. [22]

Following a personal inspection by Washington on July 21 and the establishment of the general hospital plan on July 27, the Continental Congress elected Dr. Benjamin Church as director general of the newly created medical department. Soon after this, Church conferred with several Massachusetts officials regarding the appointment of apothecaries for the medical store at Watertown. On August 3, a committee of the Provincial Congress advised "that the Medical Store in Watertown be continued where it now is, and that Mr. Andrew Craigie, appointed by the late Congress Apothecary to the Colony, be directed to take charge thereof, and prepare the necessary compositions; and that Mr. James Miller Church be appointed Assistant Apothecary to put up and distribute said Medicines...." [23]

The medical supplies were slow in coming from Philadelphia, as we have already noted. On the other hand, troops were arriving daily, placing an increased demand on all types of supplies, including drugs. One event which undoubtedly resulted in delays in establishing proper supply depots was the startling discovery that Director General Church was guilty of holding treasonable correspondence with the enemy. On October 16, Congress elected Dr. John Morgan to replace Church. [24]

On December 2, by order of Morgan, Apothecary Craigie made an inventory of the medical supplies in the general hospital at Cambridge. The inventory included 120 different items, but only limited quantities of the essential drugs. [25] There were 52 pounds of Jesuits' bark, 18 pounds of cream of tartar, 76 pounds of purging salts, 1 pound of camphor, 5 pounds of jalap, 1 pound of ipecac, and ½ pound of tartar emetic. The 44 pounds of gum ammoniac was reported "damaged," and the 86 pounds of rhubarb was described as "bad." [26] An inventory of medicines held by the different regimental surgeons in Massachusetts indicated that all regiments had "but few medicines" except for Colonel Hand's, which reported "a good supply." [27]

However, this rather meager inventory of drugs probably was not inadequate. The siege of Boston resulted in few wounded soldiers, and there was a surprisingly small amount of sickness in the army during the winter of 1775-76; furthermore, towns not too distant still had a limited supply of drugs on hand. Smith and Coit, of Hartford, Connecticut, informed "their good Customers, and the public in general, that notwithstanding the entire stop to Importation which hath long since taken place, they still have on hand, small Quantities of most Articles of the Apothecary Way ... which they mean to sell at a reasonable retailing Price." [28] Jacob Isaacks of Newport, Rhode Island, similarly advertised "a complete assortment of genuine Medicines, with furniture for containing the same, to the amount of about 300 pounds sterling; which medicines were purchased with cash, and will be sold, at the prime cost and charges, without any advance. Any of the lawful or Continental bills now current will be taken in pay for the above medicines." [29]

Drug supplies also were quite adequate in Boston during the British occupation. Sylvester Gardiner at "The Sign of the Unicorn and Mortar in Marlborow Street" reported that "all kinds of the best and [Pg 114] freshest drugs and medicines ... are continued to be sold as usual." However a cautionary note was added that drugs and medicines had been "constantly imported every fall and spring to June last." Implicit in the advertising is the suggestion that the securing of new supplies was highly uncertain. [30]

A letter dated December 2, 1775, from a British officer in Boston to a friend in Edinburgh observed that "many of our men are sick, and fresh provisions very dear." However, the officer added, "but the Rebels must be in a much worse condition...." [31] Drugs were imported into Boston during the siege as evidenced by an advertisement on February 22, 1776, announcing "just imported from

LONDON and to be sold at Mr. Dalton's Store, on the Long-Wharf, a proper assortment of Drugs and Medicines of the Best quality in Cases." [32]

By the end of February 1776, Washington had decided to try to end the siege of Boston by seizing Dorchester Heights and placing his artillery there in a position to bombard the town. General Howe believed it was time to leave, and the British evacuated on March 17.

As the Continental Army moved into Boston, there was an outcry that the British had poisoned a supply of drugs left behind. On April 15 the *Boston Gazette* reported that "it is absolutely fact that the Doctors of the diabolical ministerial butcher when they evacuated Boston, intermixed and left 26 weight of Arsenick with the medicines which they left in the Alms House." [33] Then, a week later, on April 22, appeared a series of testimonials that had been made by Joseph Warren, Daniel Scott, and Frederick Ridgley at Watertown on April 3d "by order of the Director-General of the Continental Hospital." Warren swore under oath that on or about March 29 he had gone into the workhouse [almshouse] "lately improved as an hospital by the British troops stationed in said town" and upon examining the state of "a large quantity of Medicine" left in the medicinal storeroom had found about 12 or 14 pounds of arsenic intermixed with the drugs, which were found "to be chiefly capital articles and those most generally in demand." [34]

Despite this incident, we have the word of Morgan that "a large, though unassorted stock of medicines" was collected in Boston when the British evacuated. [35] Hospital Surgeons Ebenezer Crosby and Frederick Ridgley reported that "at the evacuation of Boston ... all the Mates of the Hospital that could be spared from Cambridge ... were employed in packing up and sending off [to Cambridge] drugs, medicines and other hospital stores, collected by order of Dr. Morgan, the quantity of which appeared great." [36]

Inasmuch as few medicines were listed in the inventory of stores left by the British on the wharfs and in the scuttled ships in the harbor, [37] it appears that most of these drugs obtained in Boston were confiscated from the homes, offices, and shops of the Loyalists who fled when the British evacuated. Morgan reported that he had taken possession of the medicines and furniture of Dr. Sylvester Gardiner's shop, and a small stock of drugs from the office of Dr. William Perkins, a private practitioner. [38] No inventory of these supplies has been located thus far, but a contemporary biographer of Sylvester Gardiner records that the confiscated drugs from his shop "filled from 20 to 25 wagons." [39] This is not unlikely because Gardiner's apothecary shop was one of the largest and most prosperous in the Colonies prior to the Revolution. [40]

Soon after the British evacuated Boston, the Greenleaf apothecary shop in Boston was again supplying medicines to the Continental Army. The Greenleaf ledger [41] shows that on May 25 the shop sold nearly £4 worth of "Sundry Medicines ... [to] the Committee of War, State of Massachusetts Bay." Then, on June 20, the Massachusetts Assembly resolved that "Dr. John Greenleaf of Boston be requested to supply the Chief Surgeon of ... Colonels Marshall's, Whitney's and Craft's Regiments [Pg 115] ... with medicines as may be necessary...." [42] A short time later the Assembly advanced "up to £50 to Greenleaf for purchasing such medicines as he cannot supply from his own store." [43]

The Greenleaf ledger shows that over £32 worth of medicines were sold for Colonel Whitney's regiment and over £36 worth for Colonel Marshall's regiment between June 13 and November 20, 1776. Thus, drugs were available; but until the fall of '76, Greenleaf was having difficulty in obtaining an abundant supply.

From Bad to Worse

General Washington, correctly foretelling that New York City would be the next British objective, marched there from Boston with as much of his army as could be induced to stay under the colors. Had it not been for the presence of Washington's forces in New York, that colony would certainly have remained Loyalist; as it was, the Patriot committees had the greatest difficulty in keeping the Tories quiet by strong-arm methods. [44]

The availability of drugs in New York prior to the arrival of Washington's forces did not seem to be particularly affected by the war. Thomas Attwood "at his store in Dock-Street" offered for sale a wide assortment of drugs and medicines, [45] while William Stewart offered "a fresh supply of Genuine Drugs and Medicines ... on the most reasonable terms either for cash or at the usual credit." [46] The citizens of New York did not even have to do without their popular English patent medicines. [47]

Washington, however, had to provide for his own medical supplies in New York. In a letter dated April 3 he ordered Director General Morgan to remove the general hospital to New York with "all convenient speed...." [48] The fixing and completing of the regimental chests was to be deferred until Morgan arrived at New York.

Morgan remained behind in Boston for another six weeks collecting medicines, furniture, and hospital stores worth thousands of pounds. "The like quantity ... could not be procured," so Morgan later claimed, "in any [other] part of America." He was also able to purchase drugs from Salem, Newport, and Norwich, and before departing for New York he completed a medicine chest for each of the five regiments at Boston, Salem, and Marblehead, as ordered by Washington. [49]

Morgan arrived in New York about June 3 and purchased some additional drugs there. By June 17 his staff had made up 30 medicine chests for the regiments at New York as well as for "the branches of the General Hospital at New-York, in the bowry and neighborhood and at Long-Island." But the number of regiments requiring medical supplies exceeded Morgan's expectations, particularly since he had been advised that "the Southward regiments" would be supplied by Congress in Philadelphia. [50]

By the middle of June, Morgan must have realized that the supply of drugs available was inadequate despite the sizable quantity brought from Boston and the small stock he was able to obtain in New York. It appears that many of the New York druggists were Loyalists, and somehow they and their stock of drugs disappeared when needed by Washington's army. For example, druggist Thomas Attwood "removed his store consisting of a general assortment of Drugs and Medicines" to Newark in May only to reappear in New York again under British occupation with a good stock of "Drugs and Medicines." [51]

The New York Committee of Safety had attempted to develop a stock of drugs early in the year when they were plentiful, [52] but in

June this supply was valued at only £30. Even this small stock was not available to Morgan because when he asked permission to purchase the medicines at "a reasonable price ... for use of the Continental Hospital" the New York Provincial Congress rejected his plea on June 26 with the explanation that this medicine was to be "reserved [Pg 116] for the use of the poor and other inhabitants of this city." [53]

With increasing demands to supply the troops in the Northern Department, Morgan turned to Philadelphia and the Continental Congress. Morgan owned a small stock of drugs in Philadelphia, and knew of another supply in the possession of the firm of Delaney and Smith, [54] so he sent Dr. Barnabus Binney to Philadelphia to forward "with all dispatch" what medicines he had there and whatever could be obtained from Congress. [55] Congress resolved on July 17 "to purchase the Medicines (now in Phila) belonging to Doctor Morgan," [56] but for nearly a month Binney was unable to obtain any additional supplies either from Congress or from private sources.

On June 25 Morgan wrote to Samuel Adams asking for power "to demand a proportion of the Continental medicines left in care of Messrs. Delaney & Smith," and he repeated the request in July. However, Morgan's only reply from Adams, dated August 5, made no mention of the Delaney and Smith drug stock. Instead Adams wrote only: "I have received several letters from you, which I should have sooner acknowledged, if I could only have found leisure. I took however, the necessary steps to have what you requested effected in Congress." [57]

Finally, on August 8, Congress directed the committee for procuring medicines "to supply the director general of the Hospital with such medicines as he may want." [58] By this time, such a resolution was hardly much consolation to Morgan. Evidence of the status of the supplies in the general hospital at New York can be gleaned from an advertisement in the *New-York Gazette* of July 29 signed by Thomas Carnes, "Steward and Quarter-Master to the General Hospital":

WANTED immediately ... a large quantity of dry herbs, for baths, fomentations, &c. &c. particularly baum hysop, wormwood and mallows, for which a good price will be given. The good people of the neighboring towns, and even those who live more remote from this city, by carefully collecting and curing quantities of useful herbs will greatly promote the good of the Army, and considerably benefit themselves.

The retreat from Long Island on August 27 and the subsequent loss of New York City to the British certainly did not help the medical supply problem. Despite the fact that part of the medical stores were shipped to Stamford, Connecticut, and another stock of supplies removed to Newark, Morgan admits that "the most valuable part was still left in New-York when the enemy had effected a landing, drawn a line across the island, and were entering New-York." [59] General Knox later told how "late in the day of the 15th of September, 1776, after the enemy had beat back part of the American troops," Morgan "came over from Powles Hook in a pettiauger, and had her loaded with Hospital stores." [60] Washington personally reported on September 16 that "the retreat was effected with but little loss of Men, tho' a considerable part of our Baggage ... part of our Stores and Provisions, which we were removing, was unavoidably left in the City...." [61]

One small bundle of private drug supplies saved from the British is reported [62] by "Doct. Prime, A Refuge from Long Island," who announced the opening of a shop in Wethersfield. The newspaper advertisement reported that Prime

... has saved from the enemy a parcel of medicines, part of which he would barter for such articles as he wants, especially shop utensils of which he had unfortunately lost the most of his own....

The medical supply problem went from bad to worse as Washington's army retreated from Harlem Heights to White Plains and then finally into New Jersey. Morgan again turned to Philadelphia for drugs, but obtained "none or next to none." Instead of ten pounds of tartar emetic which Morgan requested from Philadelphia druggist Robert Bass and the newly appointed Continental Druggist, William Smith, four ounces was all that he received, but with "a proper apology." [63]

On September 21, the supply of bark was completely exhausted, and Washington was furious. On September 24 in a letter to the

President of the Congress, Washington charged that the regimental surgeons were aiming "to break up the Genl. Hospital" and that they had "in numberless Instances [Pg 117] drawn for Medicines, Stores, &c. in the most profuse and extravagent manner for private purposes." [64]

To make matters worse, new troops continued to arrive without medical supplies. For example, those from Maryland arrived at White Plains with their regimental surgeons fully expecting Morgan to supply them with medicines, even though the Maryland Convention on October 4 had ordered that these troops be supplied with medicines by the Maryland Council of Safety before their departure. [65]

Morgan thought he had at least one small but safe stock of drugs. Barnabas Binney, who was sent to Philadelphia in July for medical supplies, was successful in obtaining "a reasonable good order" about the middle of August, including "30 lb. Camphor; 10 lb. Ipecac; 7 lb. Opium; 50 lb. Quicksilver; 40 lb. Jalap; 68 lb. Manna; 186 lb. Nitre; 200 lb. Cream of Tartar; 269 lb. Bark; and other important articles." [66] However, since these supplies arrived at Newark just as Washington was beginning to pull out of Long Island, they were deposited at a newly established hospital under Cutting, the assistant apothecary. [67]

When Morgan finally began drawing on these supplies, Dr. William Shippen had been placed in charge of the hospitals in New Jersey and the medicines had been turned over to him by a vote of Congress. [68] Finally, on January 9, 1777, Congress dismissed Morgan as director general without giving any reasons except to indicate indirectly that it was due to his inability to provide adequate medical supplies. [69] To add insult to injury, on February 5 Congress asked "what is become of the medicines which Dr. Morgan took from Boston ..." and resolved to "take measures to have them secured, and applied to the use of the army." [70]

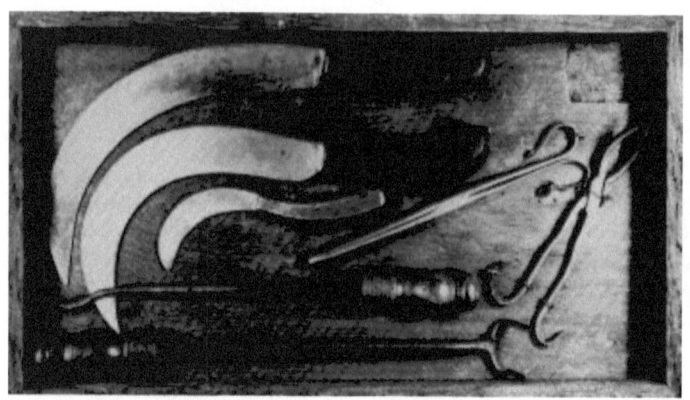

Figure 2.—Set of surgical instruments used by Dr. Benjamin Treadwell during the Revolution. Included are three amputation knives, forceps, a ball extractor, and two surgical hooks. Preserved at the Medical Museum of the Armed Forces Institute of Pathology. (*Photo courtesy of Armed Forces Institute of Pathology.*)

Meanwhile, in New York City the supply of drugs had returned to normal or near normal within a few weeks after the British occupation. On September 30, 1776, Thomas Brownejohn announced the opening "of his medicinal store at the corner of Hanover-Square ... where gentlemen of the army and navy can be supplied at the shortest notice with all kinds of medicines on the most reasonable terms." On December 16 Richard Speaight announced that he "has once again opened his Shop at the sign of the Elaboratory in Queen-Street," and a week later [Pg 118] Thomas Attwood returned from Newark to open "his store of Drugs and Medicines in Dock-Street." To touch upon the sympathy of the Loyalists, Donald McLean, "Surgeon of the late Seventy-Seventh Regiment," reported in January 1777 that he was "now happily delivered from his late captivity" and again opening a shop in Water-Street for drugs and medicines. [71]

Importations from London commenced as early as December 1776 when "the Brig Friendship lying at Beaches Wharf" offered for sale "An Assortment of Drugs, Consisting of Bark, Opium, Rhubarb, &c." In April 1777 Speaight advertised "a fresh Importation ... from the original ware-houses in London," and, in June, Attwood advertised "A large and general Assortment of Drugs and Medicines freshly imported.... Several Medicine Chests complete, fitted up in London, with printed Directions." [72]

Importation by the British was not without its problems, however. Joseph Gurney Bevan, owner of the Plough Court Pharmacy in London, wrote Dr. Traser in Jamaica on October 25, 1777:

I hope thou will be pleased with the Bark. It is very good and the best I have seen this year, but I do not think any Bark in town is equal to what I have seen in former years. Thou wilt note the snake root to be very dear. The cause is the stoppage of the American trade. Opium is also much higher than I ever knew it. The insurance is raised on account of the American privateers.

Answering a letter from William Stewart of New York, Bevan wrote on March 5, 1777:

I wish it were yet in my power to ... forward the medicines and utensils thou hast written for. But on inquiry I am informed that it is not permitted that anything shall yet be sent to New York in a mercantile way. Therefore I must defer till the wanted intercourse between us and you is re-established.... I want to advise thee to buy what snake root thou cans't pick up which I believe if sent hither at the first opening of the trade, will turn to good Account.

Bevan was still reluctant to make any shipments in April because the "ships and cargoes on their arrival at New York will be at the mercy of the persons in command there," but on September 4 he shipped a large order to McLean. [73] During the remainder of the war, the Plough Court Pharmacy continued regular shipments to McLean as well as to Stewart and to Brownejohn.

"Medicines — None"

Morgan's chaotic situation at New York was mild compared to the conditions at Fort George and Ticonderoga in the Northern Department. Dr. Samuel Stringer, medical director of the Northern Department, wrote General Washington on May 10, 1776, that the majority of the regimental surgeons had neither medicines nor instruments, and that there was no possibility of getting them in Canada. Washington replied that he would direct Dr. Morgan to send the required supplies, and ask for additional help from Congress.

[74] However, until early in June, Morgan was in no position to outfit medicine chests for any of the troops at New York, much less for the army in the north; and Congress didn't even get around to directing "the committee appointed to provide medicines ... to send a proper assortment of medicine to Canada" until June 17. [75]

After Morgan had established the general hospital at New York, he wrote to Samuel Adams on June 25 that

... the state of the Army in Canada ... for a supply of medicines is truly deplorable. General Gates sets out to-morrow to take command of the Army in Canada. Dr. Potts will accompany him. I have therefore given orders to supply him from the General Hospital with a large chest of such medicines as I can best spare, and which can be got ready to-morrow before his departure. [76]

Until July 24, the only medicines to arrive at Fort George were the "few that Dr. Potts brought with him" even though Morgan had, according to Stringer, promised to send "by the first sloop twenty half-chests of medicines" put up at New York for ten battalions in the north. Stringer therefore asked permission of General Gates at Ticonderoga to "go forth to York and see the medicines forthwith forwarded by land, until they can be safely conveyed by water." Permission was granted on July 29 and Stringer departed for New York. [77] Meanwhile, Morgan had written Potts on July 28 that he had sent Dr. James McHenry to Philadelphia for drugs, and that he was [Pg 119] sending Andrew Craigie to Fort George to "act as an Apothecary." Morgan also asked for an inventory of drugs on hand in the Northern Department. [78]

Stringer spent only a day or two in New York with Morgan—just long enough to intensify their personal feud over responsibilities and authority. Stringer determined that the "twenty half-chests" apparently were a figment of someone's imagination, because supplies in New York were almost as bad as they were in the north. Also, he learned that Morgan was sending a box of medicine northward "under the care of the Surgeon of Col. Wayne Regt." [79] that was undoubtedly intended to serve only as a regimental chest. Stringer then hurried on to Philadelphia just in time to intercept McHenry, who had obtained "an order from the Committee of Congress for 40 lb. Bark, 10 [lb.] Camphire and some other articles." [80]

Stringer wrote Potts on August 17 that at last he had obtained an order for medicines that would be packed in two days, but added "when you'll receive them God knows." He also reported that "there will also arrive another Box under the care of Doct. McHenry containing only 5 articles of which there is but 30 lbs. Bark and I think not a purgative except some few pounds of Rhubarb and a little Fol. Senae." [81] McHenry, however, only got as far as New York with his meager supplies, because Stringer discharged him from the service in an attempt to show both Morgan and Potts who had the most authority. [82]

Stringer's inexcusably long absence from his hospital post and failure to send the needed medicines so aroused General Gates that he wrote the President of the Congress on August 31 as follows: [83]

The Director of the General Hospital in this department, Doctor Stringer, was sent to New-York three and thirty days ago, with positive orders to return the instant he had provided the drugs and medicines so much wanted. Since then, repeated letters have been wrote to New-York and Philadelphia, setting forth in the strongest terms the pressing necessity of an immediate supply of these articles.

Finally, almost a month after his arrival in Philadelphia, Stringer set out for Albany with a small stock of drugs. On September 7 he wrote Potts from Albany that he hoped the small supply that he obtained and the chest of medicines that Morgan had just sent would hold out until he could obtain additional supplies in New England, where he was then headed "to ransack that Country of those articles we want." [84]

Meanwhile, Potts at Fort George had started making the desired inventory of medicines. It came as no surprise to anyone that the situation was deplorable—indeed, it was worse than that. On August 31 a committee of surgeons at Ticonderoga prepared at General Gates' order "A Catalogue of Medicines Most Necessary for the Army." This list, undoubtedly representing the minimum requirements of each battalion, called for 20 pounds of bark, 4 pounds of gum camphor, 2 pounds of gum opium, 3 pounds of powdered ipecac, 4 pounds of powdered jalap, 2 pounds of powdered rhubarb, 15 pounds of Epsom salts, and 3 pounds of tartar emetic among two dozen different medicines. [85] Instead of these minimum requirements, regimental surgeons at Ticonderoga, Crown Point, Mount Independence, and Fort George presented inventories (mostly dated September 8) that clearly emphasized their destitute condition.

The first New Jersey battalion at Ticonderoga reported "No Jallap, Rhubarb, Salts, or Ipecac"; while Colonel Whilocks' regiment at Ticonderoga reported "No medicines exclusive of private property." The five companies of artillery at Fort George reported "Medicines—None," as did the 24th Regiment at Mount Independence. Others reported small or "tollerable" assortments of medicine. A close examination of the inventory of the Pennsylvania 6th Battalion at Crown Point shows it to have been [Pg 120] lacking bark, ipecac, rhubarb, camphor, and salts; and only one-half ounce of jalap and 2 ounces of gum opium remained in the chest outfitted by Christopher and Charles Marshall on April 25 in Philadelphia. The 15th Regiment of Foot at Mount Independence claimed 2 ounces of bark and 1½ ounces of gum opium, while the 6th Regiment at Ticonderoga was as well off as any with one-half pound of bark and 4 ounces of gum opium. [86] Compared with the minimum need of 20 pounds of bark and 2 pounds of gum opium, even this was not of much comfort.

The inventory "of the Medicines in the Continental Store at Fort George" dated September 9 was not very comforting either. While the store included 137 different items, including equipment and containers of all the capital medicines, only Epsom salts appeared to be available in a sufficient quantity. Seven pounds of rhubarb were also on hand, but conspicuous by their absence were bark, ipecac, jalap, gum camphor, and gum opium. [87]

With their continuous requests and demands, the regimental surgeons made life miserable for Potts. Surgeon Mate of the Pennsylvania 1st wrote that the "Chest of Medicine ... is not yet arrived but expect it hourly...." Trumbull asked: "Have your Medicines arriv'd? Have Stringer or McHenry made their appearance yet? Our people fall sick by Dozens. I not a Pennys worth of Medicine have for them, even in the most virulent disorders." Surgeon Johnston begged: "Pray if possible send me 4 pounds Pulv. Cort. Peruv. [Bark] and 3 ounces Tart[ar] Emet[ic]. With those medicines I think I could restore a number of our best Men to perfect Health." [88]

In those instances where some drugs were on hand, the shortage of pharmaceutical equipment hampered, if not prevented, the preparation of proper dosage forms. Surgeon McCrea on board the *Royal Savage* wrote on September 2 that he "found a great inconvenience for want of scales & waits," [89] and the surgeon at Crown Point wrote on September 19 that "the Medicines which I rec'd a few days ago will be of very little Benefit as I have no fit Mortar &c. to prepare them with & must use them in Decoction." [90]

It wasn't until October that any relief arrived, and even then there were disappointments. Andrew Craigie, at Fort George, received a wagonload of herbs on October 3, but, as Craigie reported to Potts, "one half the load is entirely useless, containing Saffron, Pink flower, and whole H[eade]d Pennyroyal, &c. &c. Dr. Brown thinks his broad shoulders would carry all the articles that are worth anything." Craigie recommended to Potts that payment should not be made for all the useless articles. [91]

The long-lost Stringer finally arrived at Albany from Boston on October 5 and reported to Gates that he had met the greatest success in procuring £5,000 of medicines. [92] Ten days later, Stringer wrote Potts that he was now forwarding "by waggon two Barrels & 1 Box of Medicines ... [which] will suffice for the present, not thinking it prudent to send up the whole, especially as we can always get them up as they are wanted." [93]

Even after the long delay, most of the supplies were still held in Albany instead of being distributed among the surgeons who needed them. This infuriated Potts to a point that even Stringer found it necessary, on October 25, to explain:

I received yesterday a letter from you ... before this time you will have rec'd such of the articles you desired as we had to spare [from] the Medicines I purchased at Boston ... I thought [it] not proper to risque [them] up here; neither were any of them in powder, and all that were so at this place we sent you, and have two hands busy in preparing more for our own use. I hope that [the shipment] sent will be sufficient for your purpose. [94]

Andrew Craigie had sent three barrels and four boxes of supplies to Ticonderoga on October 22, [95] but the shipment obviously did not suffice. On November 7 Stringer wrote that "as soon as possible the Medicines you wrote for shall be prepared and sent, but they are chiefly to be pulverized." In his typical style he added, "I cannot conceive what use you will have for five sieves when you have no large mortar." [96]

The November 27 report of the committee of Congress on the conditions in the general hospital [Pg 121] at Fort George indicates that the supply situation was at last reasonably good, [97] but by this time the season was far advanced and the forces had to retire to winter quarters. Stringer was relieved of his command along with Morgan early the following year. Unlike that of Morgan, Stringer's dismissal appears to have been based on reasonably good grounds.

Privateers to the Rescue

Despite Congress' slow start in providing medical supplies, its members realized as early as December 1775 that additional sources of supply outside the Colonies would be required. On December 23 they heard that £2,000 of medicines, surgeon's instruments, and lint and bandages were required by the army, and on January 3, 1776, the Secret Committee reported to Congress that these supplies should be imported as soon as possible. [98]

In September 1775 Congress had created the Secret Committee to supervise the export and import of vital materials required for the war. Licenses to leave port were given shipmasters on the condition that they would return with vital military stores. Under this dispensation, American ships set out for Europe, Africa, and the West Indies in search of essential supplies. [99] Many months were required, however, to establish such importation as a significant source of supply, and this was especially true with regard to medical supplies.

The delay in initiating importation can hardly be charged as the only or even the main reason for medical supply shortages in 1776. For example, in August of that year, when at least a half-dozen medical supply officers were pleading for drugs from Congress in Philadelphia, John Thomson of Petersburg, Virginia, advertised that he had for sale "Rhubarb and Jalap, Glauber and Epsom Salts, Jesuits Bark" and a host of other supplies. [100] Whether or not Thomson's supplies constituted any significant amount, the very fact that he had to advertise them indicates a lack of coordination and communication between those urgently seeking supplies and those selling them.

Even more frustrating were those suppliers right under Congress's nose advertising essential drugs. Suppliers like Dr. Anthony Yeldall at "his Medicinal Ware-House" were still advertising "Bark, Camphire, Rhubarb, &c" in July of '76. [101] Philadelphia was second only to New York for Loyalists, and Yeldall was later proven to be a strong Tory. Then there were those who were neither Patriot nor Loyalist; they were just indifferent to the cause for American independence, and thus insisted on cash, even though six months' credit was the common practice just prior to the war. In 1771 in Philadelphia one druggist regularly gave a 15 percent discount on all purchases if paid within six months and 7½ percent discount was allowed for payments between six and nine months, but interest was expected on all debts over a year's standing. [102]

The business-minded members of Congress tried to follow prewar methods by seeking credit. Merchants who sold on credit found that, when they finally were paid, they received paper money backed only by a promise to exchange for gold and silver at some future time. Furthermore, they were caught in a spiraling inflation, and often found that when they finally received their money from Congress it then would cost them twice as much to replenish their stocks. Medical supply officers therefore found it necessary to pay ready cash for merchandise out of their own pocket, and sometimes they had to wait six months for reimbursement from Congress.

As we have noted, by the fall of 1776 Boston had become a better source of supply of drugs than Philadelphia, although it had been occupied by the British for nine months and Morgan had removed most of the drugs left there the previous May. This was primarily due to a single factor—the American privateer. British shipping was vulnerable to the American privateers, which were fast vessels well suited to this kind of enterprise. Well over 1,000 captures were made during the war by Massachusetts privateers alone, and the arrivals of rich prize ships at New England ports became frequent. [103]

The Greenleaf ledger confirms that drugs were included in some of these prize ships. On December 14, 1776, Greenleaf records the receipt of £62 from the Massachusetts government in payment for "an invoice of Druggs taken from the prize ship Julius Caesar." Greenleaf received an even larger stock "of druggs taken in the prize Brig Three Friends" in [Pg 122] March 1777. This was valued at over £170, and was also used by Massachusetts to pay on its account with Greenleaf, largely for outfitting its privateers. [104]

On June 30, 1777, J. G. Frazer of Boston wrote Dr. Potts, still at Ticonderoga, as follows: [105]

I have the pleasure to give you this Early notice of a prize ship being sent into Casco Bay last week with four tons of Jesuits Bark on board for one valuable article besides a great quantity of other stores for the British Army at New-York.

Brisk Business in Boston

A series of letters to Director General Potts from Apothecary Andrew Craigie, who was on a purchasing trip through New England, gives us an interesting glimpse into the situation. On August 29, 1777, Craigie wrote Potts from Springfield [106] that he had just arrived from Wethersfield where he purchased 222 pounds of bark of excellent quality. He saw it weighed and repacked, and left the necessary instructions for shipment to Albany. Having heard that "a quantity of Bark & other articles are arrived at some eastern ports" Craigie took off for Boston where he wrote Potts on September 1 as follows: [107]

I wrote you from Springfield aquainting you that I had engaged 222 lb. Bark at the Price [£5 per pound] Mr. Livingston mentioned to you; it being very dear induced me to engage a less quantity than you proposed 'til I should make enquiry here. I find to my great mortification that it is 40/[shillings] less than that in Wethersfield. I wish we could get clear of that engagement, and at least think some adjustment should be made as I am informed it cost Mr. Livingston who bought it at publick sale only 3 Pounds at which price I expect to engage 1 or 200 lb. tomorrow.... In the morning I go to Cape Anne about 40 miles from this, after medicines that have lately arrived....

Recalling Stringer's long absence of the previous year, Craigie concluded:

I shall pay particular attention to, and if to be had, procure the articles, but everything is very dear. I hope not to exceed the time you have limited.

Craigie returned to Albany on September 20 and advised Potts that he "succeeded in procuring medicines as expected" and that he had "on the road 2 covered waggons of capital medicines &c." [108] The shipment included 200 pounds of bark that Craigie bought at £3 a pound, and waiting for him in Albany were also the 222 pounds of bark, for which he was billed at £5 a pound plus £23/10 "Carting and Expenses." [109] Payment had not been made by November 10, [110] nor was there any evidence of an adjustment.

At the same time that Craigie was in Boston purchasing supplies for the Northern Department, Apothecary Jonathan B. Cutting of the Middle Department was also there, competing with him. [111] Furthermore, several agents for the Congress (Thomas Cushing, Daniel Tillinghast, and John Bradford) were purchasing drugs for the Continental Navy. Greenleaf's ledger records that between January 23 and May 28 over £500 worth of medicine chests and sundry medicines were sold to "The United American States" for the Continental frigates *Boston, Hancock, Providence,* and *Columbus.*

This competition among various branches of the army and navy led to a brisk business in Boston. Druggists in nearby communities chanced the British blockade to send supplies which they had on hand. For example, Jonathan Waldo, an apothecary at Salem, Massachusetts, recorded in his account book [112] on April 8, 1777, that "13 packages and 4 cases of medicines are ship'd on Board the Sloop called the Two Brothers Saml West Master. An Account and [illegible word] of Mr. Oliver Smith of Boston Apothecary and to him consigned." Evidence of the war appears in the footnote to the entry, however. It reads: "The cases are unmarked being ship'd at Night. Error Excepted. Jon. Waldo."

The Situation Improves

Oliver Smith, advertising in a Boston newspaper in October 1777, clearly emphasized the fact that "A Large and Valuable Assortment of Drugs and Medicines" were on hand. Included in the listing were bark, gum camphor, gum opium, jalap, rhubarb, and salts. [113]

Back in Philadelphia, the supply situation was also improving. William Smith, Continental Druggists, received over $5,000 from Congress for drug purchases, [114] [Pg 123] and the Marshalls also continued to furnish Congress with a variety of medical supplies in amounts upwards of $4,000. [115] Drugs were occasionally being imported into Philadelphia despite the British blockade. In January 1777, Robert Bass, an apothecary in Market Street, advertised [116] "A Quantity of Peruvian Bark, just imported ... together with Drugs and Medicines of most kinds." Bass was supplying the Northern Department with drugs in February 1777, but, according to a letter from John Warren to Potts, "he is determined not even to pack them untill he shall receive the money in payment for them." [117] In March, Bass wrote Potts directly that

... if in future you want any compositions let me know in time that I may have them ready. I cou'd not send a full quantity [of] fly Plasters, but am this week making a large quantity of most kinds and shall send of deficiency in your next order. [118]

In June, Christopher and Charles Marshall also received "a small assortment of valuable medicines, just imported and to be sold" [119] to replenish their stock. Even Congress purchased directly certain of the importations, on May 28, 1778, for example, ordering that "755 42/90 dollars be advanced to the Committee of Commerce, to enable them to pay Andrew and James Caldwell, the freight of sundry medicines imported in their sloop from Martinico." [120] Many of the British prize ships were carried to the French island of Martinique in the West Indies for trans-shipment of their cargoes.

These shipments however did not meet with the requirements for medical supplies. In March, Apothecary Cutting, then stationed at the "Continental Medicine Store in Fourth-Street," Philadelphia, advertised that "any price will be given for old sheets, or half worn linen proper for lint and bandages," while, in May, Commissary Hugh James advertised that "a handsome price will be given for Vials and Corks." [121] The problems of medical supplies were often brought to the attention of the public. Thomas Carnes, "Quarter Master and Steward" of the American hospital in New England, advertised in several papers that he

> is authorized to make known in this public manner, that no Expense shall be spared in future in making the most ample Provision for the sick and wounded of the Army.... Proper medicines will be prepared, not only by General Hospitals, but by Regimental Surgeons. The Difficulties the Sick and Wounded met with the last Campaign arose from the unsettled State of the Army, and the Distance Medicines, and other Necessaries used to be sent. [122]

The reorganization of the medical department by Congress, including the establishment of "two Apothecaries" and their duties, was published in the *Pennsylvania Packet* on April 15, and a front page account presenting "directions for preserving the Health of Soldiers" was featured in the next issue. [123]

Dr. Potts wrote the Medical Committee in Congress on April 3, 1777:

> I have the Honour to enclose you a Return of the Medicines & Stores belonging to the General Hospital in the Department, which I have received from Doctor Samuel Stringer, these with what I brought with me from Philadelphia & some few I expect from Boston will be quite sufficient for this campaign.

In contrast to the time when stores were short in '76, the chairman of the Medical Committee, M. Thornton, was quick to reply on April 12 that

> ... we are highly pleased with your having the prospect of a sufficient supply of medicines in your Department for the ensuing Campaign, & approve of the returns you have made us. [124]

Valley Forge

Washington's forces were defeated at Brandywine on September 11, 1777, and on September 25 the British army occupied Philadelphia. Washington, after trying without success to dislodge them by a sudden attack at Germantown on October 4, retreated to Valley Forge.

Business in Philadelphia under British occupation continued much as it had under American control, except for a few missing suppliers and a few new ones. [Pg 124] One druggist who was little in evidence after the war commenced was back in business advertising within [Pg 125] two weeks after the British occupied Philadelphia. It was William Drewet Smith (not to be confused [Pg 126] with William Smith) who advised "friends and customers ... that they can be supplied with Medicine and Drugs as usual, at his shop in Second-Street." To indicate that he was expecting an active business, Smith also advertised for "a person who can be well recommended for honesty and sobriety ... to attend a Druggist's Shop." [125]

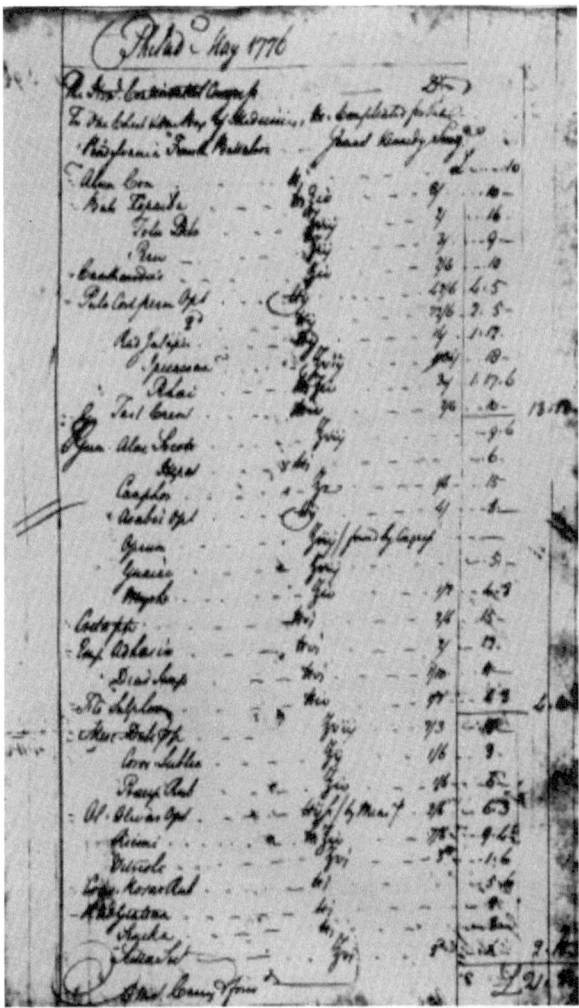

Figure 3.—Page from the Waste Book manuscript of the Christopher Marshall, Jr., and Charles Marshall apothecary shop in Philadelphia. This is the first page of the contents of a medicine chest furnished on order of the Continental Congress for the Pennsylvania 4th Battalion. Preserved at the Historical Society of Pennsylvania, in Philadelphia.

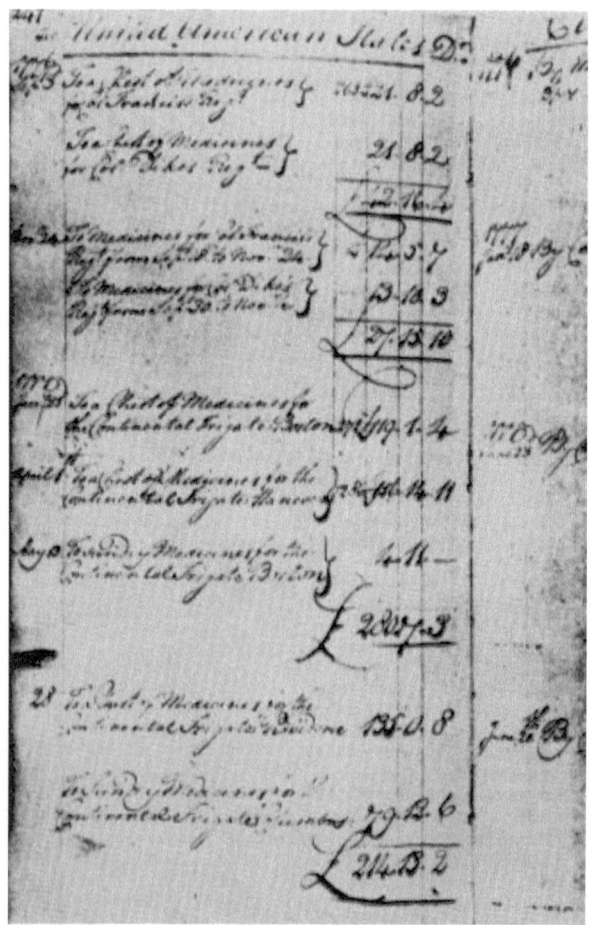

Figure 4.—Page from the ledger of the Greenleaf apothecary shop in Boston, showing the accounts between September 3, 1776, and May 28, 1777, with "the United American States" for outfitting ships of the Continental Navy. Preserved at the American Antiquarian Society, Worcester, Mass.

During the British occupation there was a large number of thefts and losses—perhaps aided by the American patriots who remained in Philadelphia—that included drugs and surgical instruments. In November an advertisement reported the loss of "a sett of Surgeons

Pocket instruments in a crimson chequered covering, with a silver clasp. Whoever will bring them to the bar of the coffee-house or to Mr. Allman, surgeons mate of the Royal Artillery, shall have a Guinea reward, and no questions asked." In April an unidentified druggist advertised: "Stolen yesterday afternoon out of an apothecary's shop Three Specie Glasses, with brass caps; one contained two pounds of native cinnabar. Whoever discovers the thief and goods shall have Twenty Shillings reward from the printer." [126]

A sign of the times is evident from the advertisement by Dr. Anthony Yeldall, who offered his "Anti-Venereal Essence at only Two Dollars." This nostrum, it was claimed, would not only cure the disease, but would "absolutely prevent catching the infection." Each bottle came with printed instructions "so that no questions need be asked." The fact that the advertisement appeared no less than 10 times from January through April speaks for its success. [127] It is interesting to note that, after the British evacuated Philadelphia, "Anthony Yeldall, Surgeon, late of the city of Philadelphia," was included among those who were charged as having "knowingly and willingly aided and assisted the enemies" and who would be brought to trial for high treason. [128]

While the British forces rested, well nourished, warm, and relatively secure in Philadelphia, Washington's troops, hardly more than 20 miles away, were tortured by cold, hunger, and disease. On December 23 there were 2,898 men at Valley Forge reported sick or unfit for duty because of lack of clothing. [129] Even so, the lack of medical supplies was nowhere near as bad as the conditions that existed in '76. Under the command of Director General Shippen and Purveyor General Potts, [130] the medical department operated a series of hospitals in such Pennsylvania communities as Easton, Bethlehem, Lancaster, Ephrata, and Lititz. The principal hospital for Valley Forge was established 10 miles away at Yellow Springs (now Chester Springs).

The largest drain on medical supplies appears not to have been during the height of winter but rather in the early spring when the medicine chests of various regiments and hospitals were being restocked for the expected spring offensive. The first step was to supplement the supply of medical supplies on hand. In late February or early March, Dr. William Brown sent Purveyor General Potts a list of needs of the entire medical department that included £20,000 worth of medicines, vials, corks, etc. [131] Dr. Brown supplemented this list with a letter to Potts dated March 11 in which he itemized the following equipment: [132]

3 doz. Boxes Small Apothecary's Weights & Scales

3 doz. Bolus knives

3 doz. Pot Spathulae

2 doz. Marble Mortars, of one pint, & Pestles

2 doz. Setts Measures, from ½ ounce to 1 [pint?]

6 doz. Earthen Vessels (deep) with handles—of different sizes, from 2 quarts to 2 galls, for boiling Decoctions, or 2 doz. copper Do. of one gallon—for that purpose.

6 doz. Delft Ware Tiles, for mixing Boluses &c. on.

While Dr. Brown was completing his report on medical supplies, he was also concluding his compilation [Pg 127] of an emergency military hospital formulary which has become known as the *Lititz Pharmacopoeia*, so named because Brown was making Lititz his headquarters at the time. The preface is dated "Lititz, March 12, 1778." The actual title (translated from Latin) reads: "Formulary of simple and yet efficacious remedies for the use of the military hospital, belonging to the army of the Federated States of America. Especially adapted to our poverty and straitened circumstances, caused by the ferocious inhumanity of the enemy, and the cruel war unexpectedly brought upon our fatherland." This formulary was published by Styner & Cist of Philadelphia in 1778, which means that it was not actually printed until sometime after June 18, when the British evacuated Philadelphia.

In the preface Brown explained that there were two types of formulas contained in the *Lititz Pharmacopoeia*; one was the "medicaments which must be prepared and compounded in a general laboratory; the others are to be mixed, as needed, in our hospital dispensaries."

The main store of drugs was housed at Manheim until late March, when Shippen ordered Apothecary Cutting to pack the medical stores there and proceed on to Yellow Springs. [133] Cutting wrote Potts on March 30 that

... the articles that we have in store are now ready to put on board the waggons excepting the want of cases to contain them.... Paper, Twine, Square Snuff Bottles & Corks are so essentially necessary to take with us, to fit up the Regimental Chests that I wish your order to buy them at Lancaster immediately. I never heard what place in the vicinity of Camp has been chosen for our temporary Medicine Shop, nor what quantities the Regimental Surgeons are to be supply'd when we get there.... [134]

On April 16 Cutting [135] wrote that the

... dispensing store is open'd here [at Yellow Springs] and we have begun to supply the Regiments in Camp.... Dr. Cochran has given orders to the Division on the left to bring their Chests first, and we propose going through the whole Army in the order in which they lay.... The best method I can think of is to act immediately about preparing new Chests upon the Northern Plan at some convenient place for all such Battallions as did not get chests from Dr. Craigie [in the] last campaign. When these new parcels are ready, let us call all the large chests into the Stores ... which are too compleat & capacious for Field Service, & in lieu of them give out our smaller ones. By this exchange, the Genl. Hospital will be well supplied with standing Chests & acquire a great variety of useful articles which are not essential in Camp.

Apothecary Cutting was concerned, however, over supplies and

... very apprehensive that the several Hospitals in this vicinity will render a further reinforcement necessary before we shall be able to compleat the whole.... To give only a few of the Capitals to each will be a work of Time, & a much more intensive piece of business than I at first imagined.

Meanwhile, Potts had sent Apothecary Craigie to Baltimore to obtain a fresh stock of drugs, and probably to prevent further friction between Craigie and Cutting. This feud started early in 1777 when Apothecary Cutting, serving with Shippen in Philadelphia, was named, over his preceptor Craigie, to head the newly organized

"Apothecary department" of the army. [136] On March 27 Craigie wrote from Annapolis advising Potts that he had been in Baltimore

... not long since and waited on Messrs. Lux & Bowly. The medicines were not come to hand but were expected.... I have engaged the whole invoice which contains several important medicines not mentioned in your list. I think the prices are full high, tho' somewhat less than Dr. Shippen affixed, and it was not in my power to procure them at a cheaper rate. They were offered £20 per lb. for all the Cantharides and much higher price for the Bark. They are not yet arrived from some place in Virginia where they were first landed. I shall examine them immediately on their arrival, and if good forward them on to Manheim, if they prove not good shall reject them, as the engagement is conditional. [137]

Then on April 4, Craigie wrote from Chester Town: [138]

I this day received a letter from Messrs. Lux & Bowley informing me, the waggons were arrived, but to their great surprise with only two packages of medicines, the others being seized near Williamsburg for the use of Virginia State. Those arrived contain but a very small share of any of the articles mentioned in your list and I believe none of the Bark and Cantharides. I shall immediately proceed to Baltimore and examine those two packages & if good send them on to Manheim, provided the price is agreeable.... I shall inquire into the circumstances of the seizure and endeavor to find out if there has been any unfair play which I can hardly suspect from the character of the Gentlemen.

[Pg 128]

Just prior to May 1, Craigie returned to Carlisle, where the "Elaboratory and Stores for the reception of the medicines &c. belonging to the military hospitals" was established, [139] and complained that he did not find the medicinal store in the order which he expected to find it:

We have many important medicines but by no means an assortment sufficient for the Army. I speak only of what is now in store. There are Medicines in different places of which I have no list.

Craigie further noted that Cutting had come up from Yellow Springs on May 1 to confer regarding plans for completing medicine chests, and would leave the following day for Baltimore where he obviously was going to try to purchase more drugs.

Craigie was puzzled by the establishment of a dispensing store at Yellow Springs, and asked whether or not the plan was

... to have the principle Store at Carlisle, where all the medicines shall be prepared, and the Chests compleated supposing the Genl. Hospitals will be more collected, and the number lessened. I would propose that an Apothecary attend each with a compleate Chest of Medicines; that the Surgeon & Physician Genl of the Army be attended by an Apothecary with good Chest, and the Regiments supply'd upon the Northern Plan. I would have an Issuing Store established at a convenient distance from the Army, from which the Hospital and Regimental Chests might occasionally be replenished. [140]

A sizable stock of drugs was finally received from Baltimore, [141] and a fairly good stock was brought down from the stores in the Northern Department, which were left well supplied by Craigie and Potts. [142] An improved plan for obtaining lint from the Moravian Sisters at Bethlehem and Lititz was proposed by Dr. Brown, [143] and "the propriety of setting the glass works at Manheim agoing" was offered as a solution by Craigie for obtaining much needed vials. [144] Local manufacturing at Carlisle [145] and "in the Jersies" [146] was used as a source of volatile and purging salts.

Gibson records [147] that between April 19 and May 3, 1778, the commands of Generals Patterson, Leonard, Poor, Glover, Scott, and Woodward turned in their medicine chests to Apothecary Cutting at Yellow Springs, and that every regiment received a standardized field box containing a definite list and quantity of necessary drugs and supplies. However, it appears likely that the project started by Cutting and continued by Craigie was not completed until late June at the earliest. [148] The "invoice of those things thought essential for the protection and health of soldiers in the field or camp" presented by Gibson [149] is actually an "Invoice of a Chest of medicines &c. compleated in the medicinal Store, N[orthern] D[epartmen]t for Thos. Tillotson Esq." [150] Inasmuch as the plan used in the Northern Department was employed by both Craigie and Cutting, the items on this invoice may serve as a reasonably good picture of the medicine chests of '78 as compared with those of '76 (see page 130).

One of the reasons for better supplies at a time when other conditions were even worse than they were in 1776 is the fact that Congress was advancing sizable, if not always completely adequate, amounts of money for the cash purchase of supplies instead of seeking credit or expecting those responsible to procure [Pg 129] supplies by using their personal money and waiting on Congress to reimburse them. During 1778, Congress advanced some $940,000 to Purveyor General Potts alone for the exclusive use of the hospital department, and these funds were in turn distributed to the proper medical procurement officers, including the apothecaries. It is significant to compare the sum of $1,095,000 provided by Congress in 1778 with £10,000 (about $27,000) which, according to Morgan, was the limit for medical and hospital supplies in 1776. [151] True, inflation had set in by 1778, and the value of money had declined greatly. For example, cantharides purchased from the Marshalls' apothecary shop in Philadelphia in 1776 cost 10 shillings per pound as compared with the cantharides Craigie purchased in Baltimore in 1778 at £20 per pound. However, the worst of the inflation was yet to come. [152]

In Summary

Initially the drug supplies for the American Revolutionary Army had come from stocks largely in the hands of private druggists. However, this source of supply was totally inadequate for a war that attained such proportions as the Revolution. Even if stocks of drugs in the Colonies had been far greater than they were, there is little reason to believe that shortages would not have developed. After all, a good many of the suppliers were Loyalists, and others were indifferent to the cause of American liberty. Even the most patriotic pharmacists were faced with a complete financial suicide, caught between a spiraling inflation and a Congress that had no money and only a promise for the future.

As if all these problems were not bad enough, the internal organization of the medical department of the army was so chaotic that, even if adequate supplies were available and if the almost insurmountable problems of communications and transportation were solved, it is almost certain that shortages would have developed at least during the campaign of 1776. Add to this the fact that any retreating army is subject to loss of supplies and the reasons for the shortages become very obvious.

The encouragement which Congress, through its Secret Committee, gave to private shippers for the importation of vital war materials offered little relief in the field of medical supplies. Importation was, of course, cut off from England, and France did not directly export any quantity of medical supplies, at least until 1778. American privateers found it much more profitable to prey on British shipping than initiating trade channels with countries which prior to the Revolution were prohibited from shipping directly to the Colonies. These channels of commerce did not develop extensively until well after the Revolution.

Hence the most immediate relief from medical supply shortages was provided by the American privateers. Drug cargoes from British prize ships, many of which were en route to New York, served as a most important source of supply, particularly in 1777 and 1778.

However, even with the most adequate supplies, competition between different branches of the army and navy and the confiscation of supplies destined for Continental troops by state militias further encouraged inflationary trends.

The number of individual drugs mentioned in various inventories was considerable, as evidenced by the listing on page 130. However, of these, only about a dozen constituted the really critical shortages. Heading the list of these "capital articles" was Peruvian or Jesuits' bark, the same cinchona from which quinine was later discovered. Tons of bark were used during the Revolutionary War, and the price more than quandrupled between June 1776 and September 1777.

The most prominent group of drugs on the list of capital articles consisted of cathartics and purgatives. Jalap, ipecac, and rhubarb were the botanical favorites, while bitter purging salts (Epsom salts) and Glauber's purging salts were the chemical choices for purging. Tartar emetic (antimony and potassium tartrate) was the choice for a vomit, and cantharides (Spanish flies) was the most important ingredient of blistering plasters. Gum opium was administered for its narcotic effects, while gum camphor, nitre (saltpetre or potassium nitrate), and mercury (pure metal as well as certain salts) were employed for a variety of purposes. Lint, a form of absorbent material made by scraping or picking apart old woven material, also often was short in supply.

Equipment shortages included surgical instruments and mortar and pestles for pulverizing the crude drugs. Glass vials for holding compounded medicines were also a supply problem, especially after essential drugs were again available.

[Pg 130]

Some of the shortages were eased, if not solved, by local manufacture. Lint was produced in large quantities in the Colonies, and glass vials were manufactured in numerous glasshouses. Even local manufacture of the purging salts and nitre aided in eliminating shortages of these essential items, and at the same time initiated the first large-scale pharmaceutical manufacturing in America.

Numerous botanicals indigenous to the Colonies were widely employed in medicine of the period, and certain ones such as snakeroot (seneka), which was widely found growing in Virginia, would have been very scarce had not an adequate supply been immediately at hand. However, attempts to substitute other indigenous plants for scarce drugs like Peruvian bark were largely unsuccessful. There is no indication that hysop, wormwood, and mallows called for during the New York crisis were ever found to be suitable replacements for any of the capital articles. Wine apparently was more useful as a substitute for bark than the bark of butternut recommended by the *Lititz Pharmacopoeia*. Peruvian bark, jalap, ipecac, camphor, opium, cantharides — these are the drugs which the American army physicians wanted, and these constituted the most serious shortage problems.

The medical supply problem was placed on relatively firm ground by the summer of 1778, having been established on the principles proven in the Northern Department under the guidance of Drs. Potts and Craigie. Furthermore, the turning point in the war had been reached. Even before Washington's forces went into winter quarters at Valley Forge, Burgoyne [153] had surrendered at Saratoga, on October 17, 1777; and, before the cold bleak winter at

Valley Forge was over, the treaty of French alliance was signed on February 6, 1778. The torments at Valley Forge proved to be the birth of a new Continental Army.

The War was still a long way from being over, and a variety of problems were yet to face the Continental Army. Inflation was yet to deal its hardest blow to the supply problem, but not even this could produce the chaos of 1776. The worst of the drug supply problem was over.

Contents of Army Medicine Chests

The following listing is an example of the contents of medicine chests ordered by the Continental Congress. The chest for the Pennsylvania 4th Battalion was filled for "Samuel Kennedy Surgeon" by the pharmacy of Christopher Jr. and Charles Marshall of Philadelphia in May 1776. The medicines are listed on an invoice in the Marshalls' waste book in the possession of The Historical Society of Pennsylvania. The contents of the Northern Department chest, compiled in the Northern Department's "Medicinal Store" for "Thos. Tillotson Esq. Surgeon & Physician General to the Army," probably was filled by Andrew Craigie at Fort George in 1778. (*Italics* denote capital article; asterisk indicates that the drug is mentioned in *Lititz Pharmacopoeia*. Contemporary English names are in parentheses following the Latin listings.)

	Pennsylvania 4th Battalion Chest	Northern Department Chest
Botanicals		
Cort[ex] Peruv[ianum] (Peruvian bark; Jesuits' bark; or bark)		4 lb.
Pulv[is] Cort[icis] Peruv[iani] (Powdered Peruvian bark)	2 lb. Opt.; 2 lb. 2nd	6 lb.

Pulvis Rad[ix] Jalapii (Powdered jalap)	2 lb.	2 lb.
Pulv[is] Rad[ix] Ipecacuan[hae] (Powdered ipecac)	8 oz.	12 oz.
Pulv[is] Rad[ix] Rhaei (Powdered rhubarb)	1 lb. 4 oz.	4 lb.
Rad[ix] Rhaei (Rhubarb root)		2 lb.
*Fol[ia] Sennae (Sennae or sena)		2 lb.
*Rad[ix] Gentian[ae] (Gentian root)	1 lb.	2½ lb.
*Rad[ix] Seneka (Senega; rattlesnake root; or snake root)	1 lb.	
*Rad[ix] Scillae Sict. (Squill; or sea-onion)	6 oz.	
Cinnamomi (Cinnamon)		1 lb.

[Pg 131]

Cort[ex] Aurant[orium] (Orange peel)		3 lb.
Fl[ores] Chamom[eli] (Camomile flower)		2 oz.
Mellisa[e] Folia] (Balm)	½ lb.	
Gum[mi] Camphor[a] (Camphor; or camphire)	10 oz.	2½ lb.
Gum[mi] Opium [also] *Opii* (Opium)	8 oz.	1 lb.
Gum[mi] Arabic[um] (Gum Arabic)	2 lb. Opt.	2 lb.
Gum[mi] Aloe Socotr[ina] (Aloe; or aloes)	8 oz.	1 lb.

Gum[mi] Aloe Hepat[ica] (Aloe; or aloes)	1 lb.	
*Gum[mi] Ammon[iacum] (Gum ammoniac)		12 oz.
*Gum[mi] Guaiac[um] (Gum guaiac)	8 oz.	¾ lb.
*Gum[mi] Myrrh[ae] (Myrrh)	4 oz.	2 oz.
*Bals[amum] Capivi (Balsam of copaiba)	1 lb. 4 oz.	2 lb.
*Bals[amum] Peruvian[um] (Balsam of Peru)	3 oz.	
Bals[amum] Tolu[tanum] (Balsam of tolu)	8 oz.	
*Ol[eum] Olivar[um] (Olive oil)	2½ lb.	
*Ol[eum] Ricini (Castor oil)	1 lb. 4 oz.	2 lb.

Drugs of animal origin

*Cantharides (Spanish flies; or flies)	4 oz.	¾ lb.
*Cera Flav[a] (Yellow beeswax)	1 lb.	4 lb.
*Mel[lis] Com[munis] (Honey)	3 lb.	
Pul[vis] Oc[uli] Canc[orum] (Powdered crabs' eyes)		1 lb.
*Sperm[atis] Ceti (Spermaceti)		3 lb.

Chemicals

*Alum[en] Com[munis] or Credem (Alum or rock alum)	1 lb.

*Creta ppt [precipitated or prae-parata] (Chalk)	6 lb.	
*Pulv[is] Crem[or] Tartar[i] (Cream of tartar)	4 lb.	2 lb.
*Tart[arus] Emetic[um] (Tartar emetic)	6 oz.	½ lb.
*Sal Nitri [or] Nitrum (Nitre or saltpetre)	4 lb.	4 lb.
Sal Absinthii (Salt of wormwood)	8 oz.	
*Sal Cath[articus] Amar[us] (Epsom salts; bitter purging salts; or bitter cathartic salts)	10 lb.	
*Sal Cath[articus] Glauber[i] [or] Sal Mirabile Glauberi (Glauber's salts; Glauber's purging salts; or Glauber's wonderful salts).	10 lb.	
*Sal Tartar[isatus] (Salt of tartar)		2 lb.
*Sal Amm[oniacum] (Sal ammoniac)		½ lb. Cd.
*Merc[urius] Corros[ivus] Sublim[atus] (Corrosive sublimate of mercury)	2 oz.	2 oz.
*Merc[urius] Praecip[itatus] Rub[er] (Red precipitate of mercury)	4 oz.	2 oz.
*Merc[urius] Dulc[is] Ppt. (Calomel)	8 oz.	
Flor[es] Sulphur[is] (Flowers of sulphur)	4 lb.	2 lb.
*Ol[eum] Vitriol[um] (Oil of vitriol)	6 oz.	
Ol[eum] Tereb[inthinae] (Oil of turpentine)		1½ lb.
Tereb[inthina] Venet[ian] (Turpentine)	1 lb. 4 oz.	

*Vitriol[um] Alb[um] (White vitriol)	4 oz.	2 oz.
*Elix[ir] Vitriol[i] (Elixir of vitriol)	3 lb.	2 lb.
Vitriol[um] Rom[anum] (Roman vitriol)	4 oz.	
Sacch[arum] Saturni (Sugar of lead)	4 oz.	
Vitr[um] Antomon[ii] Cerat[um] (Cerated glass of antimony)	3 oz.	
*Extr[actum] Saturni [also] Acetum Lithargyrites (Litharge of lead; litharge vinegar; or extract of Saturn).	11 oz.	

[Pg 132]

Tinctures

*Tinc[tura] Thebaic[a] [or] Tinctura Opii [or] Laudani Liquidi (Tincture of opium; thebaic tincture; liquid laudanum; and Sydenham's laudanam).	12 oz.	2 lb.
*Tinct[ura] Myrrh[ae] & Aloes (Tincture of myrrh and aloes).	1 lb. 12 oz.	
Tinct[ura] Cinnam[omi] (Tincture of cinnamon)	2 lb.	

Spirits

Sp[iritus] Sal[is] Ammon[iaci] (Spirit of sal ammoniac)	1 lb. 5 oz.	
Sp[iritus] Nitri Dulc[is] [also] Sal[is]	2½ lb.	1 lb. 12

Vol[atilis] (Sweet spirit of nitre)		oz.
Sp[iritus] Lavend[ula] Co[mpositus] (Compound spirit of lavender)	1 lb. 4 oz.	1½ lb.
Sp[iritus] Vini Rect[ificatus] (Rectified spirit of wine)	1 lb. 4 oz.	

Miscellaneous preparations

*Cons[erva] Rosar[um] Rub[rarum] (Conserves of red roses)	1 lb.	
Conf[ectio] Cardiac[a] (Cordial confection)		1 lb.
Elect[uarium] Asthmatic[um] (Asthmatic electuary)	1 lb. 1 oz.	
*Elix[ir] Paregor[icum] (Paregoric elixir)		2 lb.
Pill[ulae] Purgant (Purgative pills)	8 oz.	
Pulv[is] e Bol[o] Compositus (Compound powder of bole with opium)		2 lb.
Linim[entum] Sapo[naceum] (Soap liniment)		3½ lb.
Sapo[nis] Venet[ian] (Venetian soap)	2 lb.	6 lb.

Ointments

*Ung[uentum] Lap[ide] Calamin[ari] (Ointment from calamine stone)	10 lb.	4 lb.
*Ung[uentum] Basilic[um] Flav[um] (Yellow basilicon ointment)	10 lb.	
*Ung[uentum] Merc[urale] Fort[is] (Strong mercurial ointment)	6 lb.	

Ung[uentum] e Gum[mi] Elemi (Ointment of gum elemi)	3 lb.	
Ung[uentum] Alb[um] Camp[horatum] (Camphorated white ointment)	3 lb.	

Plasters

*Emp[lastrum] Adhesiv[um] (Adhesive plaster)	6 lb.	
Emp[lastrum] Diach[ylon] (Simple diachylon plaster)	6 lb.	2 lb.
Emp[lastrum] Diach[ylon] c[um] G[ummi] (Diachylon plaster with gum)	1 lb.	
*Emp[lastrum] Epispast[icum] [also] Epithema Vesicatorium (Blistering plaster; vesicatory plaster).	1 lb.	
Emp[lastrum] Stomach[icum] Majest. (Stomach plaster)	1 lb.	

Surgical dressings, etc.

*_Linteum Praeparatum_ (Lint)	1 lb. fine	
Tow	12 lb. fine	
Sponge	4 oz. fine	
Twine	1 lb. fine	½ lb.
Tape	1 piece	2 pieces
Fracture pillows	2	

Splints	2 p. Sharps 34 doz.	
Thread		4 oz.
Needles		7 common
Pins		½ thousand
Compresses		6 doz.
Bandages		700
Flannel		6 yds.
Shears		2 pr.
Rags		1 bundle

[Pg 133]

Surgical instruments

Director	1	1 steel
Probe, silver	1	1
Forceps	1	
Catheters	1 silver	
Amputating instruments		1 set
Trepanning instruments	1 Trepan	1 set
Lancets	2 best crown, 4 common	
Tourniquets	1 Brass with ligatures	8 common
Syringe, pewter	4	2
Syringe, ivory	2	

48

Glyster pipe arm'd	6	
Tooth-drawing instrument	1 Crow Bill	

Pharmaceutical equipment

Scales and weights	1 box	1 set
Mortar and pestle	1 Brass, 1 Glass	
Tyles (pill tiles)	2	
Spatulas	1 wooden handle, 1 iron handle	1 large, 1 pocket
Bolus knife	1	
Plaister knife (plaster spatula)		1
Leather skins	2 lb.	

Miscellaneous supplies

Bottles	Assortment	Assortment
Gallypots	1 doz.	Assortment
Vials	6 doz. sorted	
Corks	10 doz.	
Pillboxes	1 pacg.	
Wrapp[ing] paper	4 quire	
Writing paper	1 quire	6 quire
Ink powder		2 papers
Quiles (quills)		14 hundred

[Pg 134]

U.S. Government Printing Office: 1961

For sale by the Superintendent of Documents, U.S. Government Printing Office
Washington 25, D.C. — Price 25 cents

[1] John C. Miller, *Triumph of Freedom, 1775-1783*, Boston, 1948, preface.

[2] Louis C. Duncan, *Medical Men in the American Revolution, 1775-1783*, Carlisle Barracks, Pa., 1931; William O. Owen, *The Medical Department of the United States Army during the Period of the Revolution*, New York, 1920; James E. Gibson, *Dr. Bodo Otto and the Medical Background of the American Revolution*, Springfield, Ill., 1937; James Thomas Flexner, *Doctors on Horseback*, New York, 1939.

[3] Lyman F. Kebler, "Andrew Craigie, the First Apothecary General of the United States," *Journal of the American Pharmaceutical Association*, 1928, vol. 17, pp. 63-74, 167-178; Frederick Haven Pratt, "The Craigies," *Proceedings of the Cambridge Historical Society* (1941), 1942, vol. 27, pp. 43-86; Edward Kremers and George Urdang, *A History of Pharmacy*, Philadelphia, 1951 edition, chap. 11; Edward Kremers, "The Lititz Pharmacopoeia," *The Badger Pharmacist*, nos. 22-25, June-December 1938; J. W. England, ed., *The First Century of the Philadelphia College of Pharmacy*, Philadelphia, 1922, pp. 84-94; *American Journal of Pharmacy*, 1884, vol. 56, pp. 483-491.

[4] Jonathan Potts Papers, four volumes of miscellaneous manuscripts at The Historical Society of Pennsylvania, Philadelphia (hereinafter referred to as Potts Papers).

[5] Journals of the Provincial Congress of Massachusetts Bay, quoted in Owen, *op. cit.* (footnote 2), pp. 22-23.

[6] Greenleaf Ledger, 1765-1778, at the American Antiquarian Society, Worcester, Mass. (The Greenleaf pharmacy was established by Elizabeth Greenleaf in 1726 or 1727. See J. L. Sibley, *Biographical Sketches of Graduates of Harvard University, in Cambridge, Massachusetts*, Cambridge, 1920, vol. 5, pp. 472-476; Jonathan Greenleaf, *A Genealogy of the Greenleaf Family*, New York, 1854, pp. 89, 91, 205, 207; *Boston Post-Boy* and *Boston Gazette*, November 8, 1762, obituary of Elizabeth Greenleaf.)

[7] Owen, *op. cit.* (footnote 2), p. 23.

[8] J. R. Alden, *The American Revolution*, New York, 1954 p. 23.

[9] Owen, *op. cit.* (footnote 2), pp. 12-13.

[10] *Journals of the Continental Congress, 1774-1789*, edited by Worthington C. Ford, Washington, D.C., 1905, vol. 2, p. 250. Nearly all excerpts from Ford also appear in Owen, *op. cit.* (footnote 2).

[11] *Ibid.*, vol. 3, p. 261. The Samuel Ward diary for September 23 records that "a parcel of medicines for the hospital" was "to be bought" (E. C. Burnett, *Letters of Members of the Continental Congress*, Washington, D.C., 1921, vol. 1, p. 205).

[12] Ford, *op. cit.* (footnote 10), vol. 3, p. 344.

[13] Burnett, *op. cit.* (footnote 11), vol. 1, p. 292.

[14] *Pennsylvania Ledger*, May 6, 1775. [William Smith in Philadelphia was selling drugs in 1772 (Potts Papers, vol. 1, folio 52).]

[15] *Pennsylvania Evening Post*, December 26, 1775.

[16] *Pennsylvania Packet*, September 11, 1775; *Pennsylvania Journal*, September 6, 1775; *Pennsylvania Gazette*, October 4, 1775.

[17] The Marshalls sold drugs to Sharp Delaney and William Smith in April 1776 (Marshall Waste Book, see footnote 20).

[18] E. T. Ellis, "The Story of a Very Old Philadelphia Drug Store," *American Journal of Pharmacy*, 1908, vol. 75, p. 57; England, *op. cit.* (footnote 3), pp. 348-350; Parke, Davis & Co., *A History of Pharmacy in Pictures*, undated booklet edited by George Bender.

[19] Ford, *op. cit.* (footnote 10), vol. 3, p. 442; vol. 4, pp. 188, 197.

[20] Christopher Jr. and Charles Marshall Waste Book, February 21 to July 6, 1776, at The Historical Society of Pennsylvania, Philadelphia.

[21] Ford, *op. cit.* (footnote 10), vol. 3, p. 442; vol. 4, pp. 188, 197; Burnett, *op. cit.* (footnote 11), vol. 1.

[22] Owen, *op. cit.* (footnote 2), pp. 18-19.

[23] *American Archives* ... Peter Force, ed., Washington, ser. 4, vol. 1-6, 1837-46; ser. 5, vol. 1-3, 1848-53. Ser. 4, vol. 3, p. 306.

[24] Duncan, *op. cit.* (footnote 2), pp. 62-64.

[25] *Pennsylvania Packet*, June 24, 1779.

[26] It is quite possible that the designation "bad" was a typographical error for "rad[ix]."

[27] *American Archives*, ser. 4, vol. 5, p. 115.

[28] *Connecticut Courant*, February 12, 1776.

[29] *Newport Mercury*, January 15, 1776.

[30] *Massachusetts Gazette*, September 7, 1775.

[31] *American Archives*, ser. 4, vol. 4, p. 159.

[32] *Massachusetts Gazette*, February 22, 1776.

[33] *Boston Gazette*, April 15, 1776.

[34] *Ibid.*, April 22, 1776. It is worth noting that Morgan did not think this important enough to include in his *Vindication* (see footnote 35).

[35] John Morgan, *A Vindication of His Public Character in the Station of Director-General of the Military Hospital, and Physician in Chief of the American Army; Anno, 1776*, Boston, 1777.

[36] *Pennsylvania Packet*, June 24, 1779.

[37] *American Archives*, ser. 4, vol. 5, p. 488.

[38] Morgan, *op. cit.* (footnote 35), pp. 102, 144; and *Independent Chronicle*, April 10, 1777.

[39] James Thacher, *American Medical Biography*, Boston, 1828, vol. 1, pp. 270-273.

[40] For biographies of Sylvester Gardiner see *Dictionary of American Biography*, New York, 1931, vol. 8, pp. 139-140; *Appleton's Cyclopedia of American Biography*, New York, 1887, vol. 2; H. A. Kelly and W. L. Burrage, *Dictionary of American Medical Biography*, New York, 1928, pp. 450-452; James H. Stark, *The Loyalists of Massachusetts*, Boston, 1910, pp. 313-315.

[41] Greenleaf Ledger (see footnote 6).

[42] *American Archives*, ser. 5, vol. 1, pp. 282, 284.

[43] *Ibid.*, p. 314.

[44] S. E. Morison and H. S. Commager, *The Growth of the American Republic*, New York, 1950, vol. 1, p. 210.

[45] *New-York Journal*, July 13, 1775.

[46] *Ibid.*, May 11, 1775.

[47] *New-York Gazette*, January 1 and January 29, 1776. For a history of the English patent medicines in America, see G. B. Griffenhagen and J. H. Young in *The Chemist and Druggist*, 1957, vol. 167, pp. 714-722, and in *U.S. National Museum Bulletin 218*, 1959, pp. 155-183 (Contributions from the Museum of History and Technology, Paper 10).

[48] George Washington, *The Writings of George Washington*, edited by John C. Fitzpatrick, Washington, 1931, vol. 4, pp. 464-465.

[49] Morgan, *op. cit.* (footnote 35), pp. 4, 9, 68; *Pennsylvania Packet*, June 19, 1779; and Washington, *op. cit.* (footnote 48), vol. 4, pp. 464-465.

[50] Duncan, *op. cit.* (footnote 2), p. 135; Morgan, *op. cit.* (footnote 35), p. 11.

[51] *New-York Gazette*, May 6 and December 23, 1776.

[52] *American Archives*, ser. 4, vol. 4, p. 1026.

[53] *Ibid.*, vol. 6, p. 1431.

[54] Morgan misspelled Delaney as "Delancey" in his letter of June 25 to Adams.

[55] Morgan, *op. cit.* (footnote 35), p. 128.

[56] Ford, *op. cit.* (footnote 10), vol. 5, p. 570.

[57] *American Archives*, ser. 4, vol. 6, p. 1069.

[58] Ford, *op. cit.* (footnote 10), vol. 5, p. 633.

[59] Morgan, *op. cit.* (footnote 35), p. 12.

[60] *Pennsylvania Packet*, June 26, 1779.

[61] Washington, *op. cit.* (footnote 48), vol. 6, pp. 58-59.

[62] *Connecticut Courant*, January 6, 1777.

[63] Morgan, *op. cit.* (footnote 35), pp. 13, 136, 146. William Smith was appointed Continental Druggist on August 20; see Ford, *op. cit.* (footnote 10), vol. 4, pp. 292-293.

[64] Washington, *op. cit.* (footnote 48), vol. 6, pp. 86, 113.

[65] *American Archives*, ser. 5, vol. 3, pp. 116, 837.

[66] *Pennsylvania Packet*, June 24, 1779.

[67] Morgan, *op. cit.* (footnote 35), p. 129.

[68] *Ibid.*, p. xxv. [For details of the manner in which Shippen moved in on Morgan to replace him eventually as director general, see Flexner, *op. cit.* (footnote 2), pp. 3-53.]

[69] *Ibid.*, p. xxxv; Owen, *op. cit.* (footnote 2), p. 55.

[70] Ford, *op. cit.* (footnote 10), vol. 7, p. 91.

[71] *New-York Gazette*, September 30, December 16, 23, 1776, January 20, 1777.

[72] *Ibid.*, December 9, 1776, April 28, June 9, 1777.

[73] Plough Court Pharmacy letterbook dated April 7, 1778, through December 8, 1779, in possession of Allen and Hanburys, London. See also Chapman-Huston and Ernest C. Gripps, *Through a City Archway: The Story of Allen and Hanburys, 1715-1954*, London, 1954.

[74] Duncan, *op. cit.* (footnote 2), p. 97.

[75] Owen, *op. cit.* (footnote 2), p. 39.

[76] *American Archives*, ser. 4, vol. 6, p. 1069.

[77] *American Archives*, ser. 5, vol. 1, pp. 651-652, 1114.

[78] Potts Papers, vol. 1, folio 77; Morgan to Potts, July 28, 1776.

[79] *Ibid.*, folio 89; Stringer to Potts, August 17, 1776. See also Gibson, *op. cit.* (footnote 2), pp. 108-109. Washington mentions Stringer's visit with Morgan in a letter to Gates dated August 14 (Washington, *op. cit.* footnote 48, vol. 5, pp. 433-435).

[80] *Ibid.*; McHenry to Potts, August 3, 1776. [Stringer arrived in Philadelphia on the evening of August 2.]

[81] *Ibid.*; Stringer to Potts, August 17, 1776.

[82] *Ibid.*; McHenry to Potts, August 21, 1776.

[83] *American Archives*, ser. 5, vol. 1, p. 1271. For a similarly worded letter to Egbert Benson dated August 22, see Gibson, *op. cit.* (footnote 2), p. 112.

[84] Potts Papers, vol. 1, folio 98; Stringer to Potts, September 7, 1776. Stringer arrived in Albany on September 5 (Potts Papers, vol. 1, folio 97).

[85] *American Archives*, ser. 5, vol. 1, p. 1266. Other items included "Acet. Com. six barrels; Alo. Hepta. 3 lb.; Calomel 2 lb.; Emp. Diachyl 10 lb.; Cantharid. 2 lb.; Gm. Guiac 1 lb.; Myrrh 1 lb.; Hord. Com. 100 lb.; Jerc. Precip. Rub. ½ lb.; Merc. Cor. Sublim. 1 lb.; Rad. Serpent. Virg. 3 lb.; Sal. Nit. 5 lb.; Spirit Sal. Ammo. 4 lb.; Ung. Diath. 3 lb.; Elix. Asthmat. 5 lb.; and Elix. Vitriol. 10 lb." Also included were six gross of vials and corks and three reams of wrapping paper.

[86] Potts Papers, vol. 1, folios 102-106, 108-111, 114, 119.

[87] *Ibid.*, folio 99. There was a listing for 170 pounds of "Cathart: Am" (Epsom salts). The 7 pounds of rhubarb was listed as "3 lb. Rad. Rhaei and 4 lb. Pul. Rhaei." Also on hand were 1½ pounds of "Mithridat" (opium).

[88] *Ibid.*, folios 73, 94, 124.

[89] *Ibid.*, folio 4; McCrea to Potts, September 2, 1776.

[90] *Ibid.*, folio 124; Johnston to Potts, September 19, 1776.

[91] *Ibid.*, folio 125; Craigie to Potts, October 3, 1776.

[92] *American Archives*, ser. 5, vol. 2, p. 923. Stringer also wrote Potts on October 6 to advise him of the stock (Potts Papers, vol. 1, folio 126).

[93] Potts Papers, vol. 1, folio 131; Stringer to Potts, October 15, 1776.

[94] *Ibid.*, folio 133; Stringer to Potts, October 25, 1776.

[95] *Ibid.*, folio 132; Craigie to Potts, October 22, 1776.

[96] *Ibid.*, folio 138; Stringer to Potts, November 7, 1776.

[97] Duncan, *op. cit.* (footnote 2), p. 110.

[98] Ford, *op. cit.* (footnote 10), vol. 3, p. 453, vol. 4, pp. 24-25.

[99] Miller, *op. cit.* (footnote 1), pp. 103-113.

[100] *Virginia Gazette*, August 24, 1776.

[101] *Pennsylvania Evening Post*, July 18, 1776.

[102] G. B. Griffenhagen, "The Day-Dunlap 1771 Pharmaceutical Catalogue," *American Journal of Pharmacy*, 1955, vol. 127, pp. 296-302.

[103] 103 Miller, *op. cit.* (footnote 1), pp. 110-112.

[104] Greenleaf Ledger, *op. cit.* (footnote 6).

[105] Potts Papers, vol. 2, folio 213.

[106] *Ibid.*, vol. 3, folio 305.

[107] *Ibid.*, folio 331.

[108] *Ibid.*, folio 346.

[109] *Ibid.*, folio 336.

[110] *Ibid.*, folio 369.

[111] *Ibid.*, folio 331; Craigie to Potts, September 1, 1777.

[112] Preserved at the Essex Institute, Salem, Massachusetts.

[113] *Independent Chronicle*, October 30, 1777.

[114] Ford, *op. cit.* (footnote 10), vol. 5, p. 748, vol. 7, p. 274, vol. 8, p. 538. (Smith received $2,490 on September 9, 1776, $2,952 on April 17, 1777, "for sundry medicines," and Caldwell & Co. received $666 on July 7, 1777, "for sundry medicine delivered William Smith.")

[115] *Ibid.*, vol. 7, p. 321. (Christopher and Charles Marshall received $4,151 on May 2, 1777, "for sundry medicines and chirurgical instruments supplied by them for the use of different battalions of continental forces.")

[116] *Pennsylvania Journal*, January 29, 1777.

[117] Potts Papers, vol. 2, folio 150.

[118] *Ibid.*, folio 153; Bass to Potts, March 17, 1777.

[119] *Pennsylvania Journal*, June 11, July 9, 23, 1777.

[136] *Ibid.*, vol. 2, folio 151; Tillotson to Potts, February 22, 1777. [Cutting served as Assistant Apothecary under Craigie at Cambridge and Roxbury. The feud has not been explored in any of Craigie's biographies.]

[137] *Ibid.*, vol. 4, folio 429; Craigie to Potts, March 27, 1778.

[138] *Ibid.*, folio 437; Craigie to Potts, April 4, 1778.

[139] *Ibid.*, folio 411; Potts to Gates, February 24, 1778.

[140] *Ibid.*, folio 441; Craigie to Potts, May 1, 1778.

[141] *Ibid.*, vol. 1, folios 41, 44; undated invoices from Lux & Bowly that undoubtedly were supplied during the spring or summer of 1778. Also, vol. 4, folio 476; letter from James Caldwell to Potts advising "I sent forward from Baltimore a case of medicine & five cases of Bark ... I have three cases more of Bark not yet up from Williamsburg where it arrived."

[142] *Ibid.*, vol. 4, folio 458; Craigie to Potts, May 1, 1778. Craigie advises: "Enclosed is a small List directed to Mr. Root [Israel Root or Josiah Root, both apothecaries from Connecticut] which I think may well be spared from the Northward, and are much wanted here. I wish therefore they may be ordered. Andrew Atekin our assistant there might come with them—he would make a good Hospital Apothecary." Also, vol. 4, folio 431, an undated "Invoice of Medicines &c. to be forwared for Head Quarters to Compleat ye Regimental Assortments for the Army of the United States in the Middle Department for the Campaign 1778."

[143] *Ibid.*, folio 419; Brown to Potts, March 11, 1778.

[144] *Ibid.*, folio 458; Craigie to Potts, May 1, 1778.

[145] *Ibid.*, folio 428; Cutting to Potts, March 25, 1778. Cutting notes: "as to volatile salts, I expect a fine parcel manufactured at Carlisle by tomorrow."

[146] *Ibid.*, folio 471; Craik to Potts, May 24, 1778. Dr. Craik, a regimental surgeon, advises: "I wish you could procure some Cathartic salts. The Regimental surgeons complain greatly for want of them.... You may engage any quantity at the salt works in the Jersies."

[147] Gibson, *op. cit.* (footnote 2), pp. 166-167.

[120] Ford, *op. cit.* (footnote 10), vol. 11, p. 546.

[121] *Pennsylvania Evening Post*, March 18, May 27, 1777.

[122] *Boston Gazette*, February 3, 1777; *Connecticut Courant*, April 7, 1777.

[123] *Pennsylvania Packet*, April 15, 22, 1777. This anonymous article was written by Dr. Benjamin Rush and reprinted as a pamphlet.

[124] Potts Papers, vol. 2, folios 158, 159.

[125] *Pennsylvania Ledger*, October 10, 1777; *Pennsylvania Evening Post*, October 14, 18, 1777.

[126] *Pennsylvania Evening Post*, November 1, 8, 13, 1777, April 29, 1778. (A large number of advertisements announcing thefts appeared during the British occupation.)

[127] *Pennsylvania Evening Post*, January 10 through April 20, 1778, and *Pennsylvania Ledger*, April 4, 15, 1778. [Yeldall advertised his "Anti-Venereal Essence" only once under American occupation, but at $4.00 per bottle (*Pennsylvania Evening Post*, August 26, 1777).]

[128] *Pennsylvania Evening Post*, June 25, 1777.

[129] Gibson, *op. cit.* (footnote 2), p. 149.

[130] It was in February 1778 that Dr. Potts assumed his office as purveyor general for the hospital department of the Continental Army with the duty of purchasing and distributing all supplies and medicines (*ibid.*, p. 154).

[131] Potts Papers, vol. 1, folio 24. (This apparently is the list prepared by Brown, even though it is not signed by him. The item "Medicines, Vials, Cork &c. £20,000" was added with the statement "The above enumerated articles should be purchased immediately," and both were in the handwriting of "W. Shippen, D.G." The document is undated.)

[132] *Ibid.*, vol. 4, folio 419; Brown to Potts, March 11, 1778.

[133] *Ibid.*, folio 428; Cutting to Potts, March 25, 1778.

[134] *Ibid.*, folio 432; Cutting to Potts, March 30, 1778.

[135] *Ibid.*, folio 441; Cutting to Potts, April 16, 1778.

[148] Potts Papers, vol. 4, folios 462, 467; Craik to Potts, May 2 and May 15. On May 2, Craik advises that "the medicine chests are much wanted in the Regiments. Doctr. Cutting had best have them filled up as soon as possible to prevent complaints." On May 15 Craik commented: "I am sorry Doctr. Cutting went away before the Regiment Chests were finished; there is great clamour about them tho Doctr. Layman is as busy as possible.... I hope Doctr. Craig[ie] will soon have his chests ready."

[149] Gibson, *op. cit.* (footnote 2), pp. 167-168.

[150] Potts Papers, vol. 1, folio 25, undated.

[151] Gibson, *op. cit.* (footnote 2), p. 178, and Duncan, *op. cit.* (footnote 2), pp. 115-116, 275.

[152] Miller, *op. cit.* (footnote 1), pp. 425-477.

[153] An interesting account of the medical aspects of Burgoyne's campaign is recorded by R. M. Gorssline in *Canadian Defense Quarterly*, 1929, vol. 6, pp. 356-363.

www.ingramcontent.com/pod-product-compliance
Lightning Source LLC
Chambersburg PA
CBHW030507220526
45464CB00006B/2698